Light pollution

Dr. Hemant Pathak

DEDICATION

Dedicated to Shri Sainath Maharaj the all omnipotent of world the most merciful.

CONTENTS

Foreword

Light pollution; provides a unique insight into the problems our planet faces in terms of clean environment, and what to do about it. This is the only books Written for academics, researchers and practitioners working in environmental pollution and management field, expressed comprehensive and interdisciplinary focus on the ecological issues associated with Light pollution to provide a complete picture of current environmental problem from cause to effect to solution

This book made of 10 years consistently research on environmental issues, makes it ideal source for students, teachers, industrialist, environmental experts and economists.

This book provides an essential guide to researchers, it offers: various causes of pollution; on the challenges and experiences in present scenario.

Simply explained, Light pollution is an important book bringing together diverse viewpoints from academia and environmental agencies and regulators, for all who wish to make a difference in how to plan and manage our Environmental resources.

Dr. Hemant Pathak
M.Sc. (Gold medalist), Ph. D.
Assistant Professor of Engineering Chemistry
Indira Gandhi Govt. Engineering College,
Sagar, MP, India

Glossary

Abatement	The reduction or elimination of pollution.
Act	A law
An absence of darkness	Artificial light makes experiencing natural night-time lighting conditions impossible in many parts of the country.
Aura	The hemisphere of light rising up from ground level encircling a light source or lighting array caused by low level mist and fog particles.
Candela	A unit of luminous intensity radiated in a particular direction.
Circadian rhythm	Animals and plants live by a rhythm which is attuned to our planet's 24-hour cycle. This is an inherited trait, which is passed on through the genes of a species. Humans may notice a change in their circadian rhythm when they travel by airplane between several time zones, characterized by sleepiness, lethargy, or a general sense that something is "off."
Community Ecology	The behaviors exhibited by individual animals in response to ambient illumination (orientation, disorientation) and to luminance (attraction, repulsion) influence community

interactions, of which competition and predation.

Ecosystem
An interactive system that includes the organisms of a natural community association together with their abiotic physical, chemical, and geochemical environment.

Ecological light pollution
includes chronic or periodically increased illumination, unexpected changes in illumination, and direct glare.

Efficacy
In lighting terms - the value of light obtained per unit of electrical energy input, i.e. lumens per watt.

Flicker
The periodic, often deliberate, flickering of light used for advertising and attraction seeking purposes can prove to be distracting and like glare, promote degrees of irritation, annoyance and distress.

Glare
light that shines horizontally and directly into a person's eyes. The excessive contrast between bright and dark areas in the field of view.

IDA
The International Dark-Sky Association (IDA), is an educational organization that seeks to preserve the natural night skies worldwide.

Illuminance The quantity of luminous flux incident upon a unit area, expressed as lumens per square metre or lux.

Light clutter The excessive grouping of lights, for example in roadside advertising which can prove a dangerous distraction to motorists.

Light Pollution Increasing environmental problem in the urban landscape which has well documented deleterious implications for the visual environment, human health, wildlife and flora.

Light profligacy Over-illumination which wastes energy and money

Light trespass Unwanted light, for example from adjacent properties, floodlights, security lights, streetlights that spills onto property which would otherwise be dark. It is often the result of over-lighting.

Lumen A unit of light (luminous flux) emitted froma point source of one candela intensity, usually expressed in kilolumens (kLm).

Luminance The luminous intensity (or brightness) of a surface or source expressed in terms of surface area, i.e. candelas per square metre.

Over illumination Artificial lighting that is brighter and on longer than required for a specific activity

Sky glow bright halo over urban, suburban, and some

rural areas at night due to bad outside lighting. Sky glow results from waste light that is beamed above the horizontal. It wastes energy, wastes dollars and washes out our view of a major part of the environment and our heritage the night sky.

Skylight The variable brightness value of daytime sky caused by sunlight scattered by particles of dust and vapour in the earth's atmosphere (skylight can reach values in excess of 2000 candelas per squaremetre).

Wattage The nominal load rating of a lamp (excludes any allowances for associated operating gear losses).

Light pollution

1. Introduction

Light is the part of the electromagnetic radiation spectrum that is visible to humans. The human eye can only see light in the visible spectrum and has different sensitivities to light of different wavelengths within the spectrum. The increase in the use of exterior lighting during nighttime has produced undesirable side effects known as light pollution.

While Light Pollution is one of the fastest growing & most pervasive forms of environmental pollution, most rapidly increasing alterations to the natural environment is the alteration of the ambient light levels in the night environment produced by man-made light. Improperly installed or shielded lights emit a large portion of their energy away from the premises and thus waste money per unnecessary kilowatt, per year.

The use of artificial electric lighting has increased rapidly over the last hundred years both in daytime and nighttime use allowing humans to adapt to 24 hour active society.

The study of global change must take into account this phenomenon called light pollution.

Current research suggests that light pollution can have lasting adverse effects on both human and wildlife health.

On Earth various species have different spectral sensitivities; many insects are able to detect ultraviolet light, which is electromagnetic radiation of a wavelength that is too short for the human eye to perceive.

Some species have a high sensitivity to a narrow band of radiation within their spectral range.

The term light pollution referred to the degradation of human views of the night sky (hiding stars). Many people assume artificial light provides safety and improves visibility. However, a large portion of lighting does neither. Lighting that is overused, misdirected, or otherwise obtrusive is simply pollution.

In addition to this, artificial night lighting can have adverse effects on wildlife as well as to humans. Light signal at wrong biological time can interfere with the normal behavior of both plants and animals.

Light pollution is an important and avoidable consequence of poor lighting design, often exacerbated by poor installation and maintenance. These factors result in light shining outwards and upwards into the sky where it is not wanted and where it often reflects off moisture and very fine particulate matter in the air giving rise to 'sky glow'.

2. Ecological light pollution

Light Pollution disrupts natural patterns of light and dark. It changes animal behaviors (confuses navigation, alters competition & predator prey interactions) and adversely affects physiology and reproduction.

The intensity, spectral quality, duration and periodicity of exposure to light affect the biochemistry, physiology and behavior of organisms.

In plants, the presence of light-sensitive chemicals provides the basis for photosynthetic activity. Light is also an important environmental modulator of growth rates and growth patterns for which changes can have profound consequences at the level of the individual plant.

Many micro-organisms and a wide variety of animals ranging from protozoan's to higher vertebrates perceive light, the ability may range from basic light perception to full visual imaging.

Research on insects, turtles, birds, fish, reptiles, and other wildlife species shows that light pollution alters behaviors, foraging areas, migration timing and routes, and breeding cycles. More complex animals have the ability to form images from information gleaned using their light receptor systems.

Ecological light pollution alters natural light regimes in terrestrial and aquatic ecosystems.

Light Pollution contributes to lower water quality by preventing zooplankton from feeding on algae, which grow more in light.

Plants and animals, including humans, have 24-hour biological rhythms under the control of the daily light-dark cycle. Light pollution hurts diurnal & nocturnal species by disrupting Physiological rhythms, including hormone levels, Behavior patterns like feeding, predator avoidance, courtship, migration. Prolonged exposure to artificial light prevents many trees and other plants from adjusting to seasonal variations. Reproduction, leading to population declines & secondary effects on other species Ecosystems.

Possible adverse ecological effects of artificial night lighting to plants, animals and humans .

3. Human physiology

Light Trespass from Outdoor Lighting has Indoor Effects Excessive artificial light can affect production of hormones, disrupt sleep patterns, and have other adverse effects on human health. It disrupts biological rhythms.

Despite of the availability of electric light over a century the possible adverse effects of artificial lighting are only partially understood.

Recent research on the link between light and cancer have shown that light and especially night lighting can be a public health issue.

4. Effect of Light pollution on Human Health

The 24-hour cycle (circadian clock) affects physiologic processes, brain wave patterns, hormone production, cell regulation, and other biologic activities.

Industrialized countries show higher risk of breast cancer than the least industrialized countries.

Melatonin is a hormone produced by the pineal gland and secreted at night which is known to help regulate the body's biologic clock. Melatonin levels drop precipitously in the presence of excessive artificial light.

• The circadian cycle controls from ten to fifteen percent of our genes.

• Disruption of the circadian clock is linked to several medical disorders in humans (e.g., depression, insomnia, cardiovascular disease, cancer).

• Dramatic increases in the risk of breast and prostate cancers, obesity, and early-onset diabetes have mirrored the dramatic changes in the amount and pattern of artificial light generated during the night and day in modern societies over recent decades.

• Researchers have concluded that excessive artificial light exposure early in life may contribute to an increased risk of depression & other mood disorders in humans.

• Excessive artificial light from outside at night may affect production of hormones, prevent healthy sleep, and have other adverse effects.

5. Light Pollution Measurement

Measurement of ecological light pollution often involves determination of illumination at a given place. Illumination is the amount of light incident per unit area. The measurement of light pollutions using by digital photography spectrophotometer.

6. Security and Safety

Public safety and security at night require a certain amount of illumination, and there is clear evidence that improved lighting leads to cost-effective reductions in crime. However, rather than increased surveillance and other deterrent effects, the benefits of improved lighting are usually attributable to increased community pride and confidence, both day-time and night-time crime decreasing together.

There is an almost universal misconception that lighting diminishes crime and increases security. In fact several government sponsored research projects in the United Kingdom and the United States of America have failed to demonstrate any benefit from lighting in the fight against crime. Studies that claim to show a beneficial effect have been convincingly shown to be flawed on closer analysis.

School districts across the United States have reduced the cost of vandalism by adopting a "Dark Campus" policy. All lights are turned off between 11pm and 6am. As well as reducing crime against school property they have achieved significant savings in their electricity bills.

Young women at brightly lit bus shelters at night report that they feel exposed to the unwanted gaze of passing traffic and would rather stand discretely in the shadows.

Perpetrators of crime need to see what they are doing, just like the rest of us and prefer to operate in well lit areas where they can also easily see if there is anyone watching them.

A criminal using a torch in an unlit area is more likely to arouse suspicion than someone wandering around brazenly in an over lit area.

Misplaced or over bright lighting on our roadways can distract and confuse and the glare can impair visibility and adversely affect the safe use of roads Lighting.

7. Conclusions

The understanding of the ecological effects is only partially understood at the moment, and the field of light pollution needs urgently further research.

Light pollution

There should be clearly multidisciplinary collaboration between physical scientists, engineers, medical experts, biologists, and ecologists.

In regard to environmental aspects, especially the measurement of light characteristics and the understanding of tropic and aquatic ecosystems need further research.

In regard to human physiology the research is rather active on the possible carcinogenic effect of artificial night lighting, but the knowledge on the influence of dark periods or dark pulses on human physiology is rather unknown.

A totally unshaded outdoor light shines about 50% of its light up into the sky where its serves no useful purpose. This wasted light represents also wasted energy. Replacement of such lights with a better designed fitting would allow the use of lower power lamps to give the same degree of illumination thus halving their running costs. This benefits the owner of the light while at the same time reducing the light polluting effect.

To assist developers, architects, lighting designers and local authority staff in applying a consistent approach to the provision of lighting it is recommended that local authorities document their lighting policy. Shield your outdoor lighting, Only use the light when you need it. Use timers and dimmers Shut off the lights when you can Use only enough light to get the job done. Use long wavelength light with a red or yellow tint to minimize impact. Humans have now so altered the natural patterns of light and dark that these new conditions must be afforded a more central role in research on species and ecosystems.

REFERENCES

1. Purves D, Fitzpatrick D, Augustine GJ, Katz LC, Lawrence C, LaMantia AS, McNamara JO, Mark WS, 2001, Neuroscience. 2nd edition. Sunderland. Sinauer Associates, Inc.

2. Berson, DM, Dunn FA, Takao M. 2002. Photo transduction by retinal ganglion cells that set the circadian clock. Science 295:10.

3. Stockman A, Sharpe LT. 2000. Spectral sensitivities of the middle and long wave length sensitive cones derived from measurements of observers of known genotype. Vision Res. 40:17111737.

4. The photopic sensitivity data, together with the standard CIE scotopic data. Available from: http://cvision.ucsd.edu/.

5. Kokoschka S. 1997. Das V Dilemma in der Photometrie. Proceedings of 3. Internationales Forum fur den lichttechnischen Nachswuchs, TU Ilmenau, Ilmenau.

6. Eloholma M, Ketomäki J, Halonen L. 2004. Luminances and visibility in road lighting conditions, measurements and analysis. Report 30. Helsinki University of Technology, Lighting Laboratory. 27 p.

7. Provencio I, Rodriguez IR, Jiang G, Hayes WP, Moreira RF, Rollag MD. 2000. A novel human opsin in the inner retina. J. Neurosci. 20:600–605.

8. Provencio I, Jiang Gm De Grip WJ, Hayes WP, Rollag MD. 1998. Melanopsin: An opsin in melanophores, brain, and eye. Proc. Natl. Acad. Sci. U. S. A. 95:340–345.

9. Alteration of natural light levels in the outdoor environment owing to artificial light sources, Cinzano et al. 2000, Monthly Notices of Royal Astron. Soc., 318, 64

10. Save the Night in Europe, a comprehensive source aimed not just at clear sky luminance.

http://savethenight.eu

11. Rea MS, Figueiro MG, Bullough JD, Bierman A. 2005. A model of phototransduction by the human circadian system. Brain Res Brain Res Rev. 50(2):213228.

12. Catanzaro G., Catalano F. A., 2000, in Cinzano P., ed., Measuring and Modelling Light Pollution, Mem. Soc. Astron. Ital., 71, 211

13. Cinzano P., 1994, References on Light Pollution and Related Fields version 2, Internal Report 11, Dep. of Astronomy, Padova, also on-line at http://www.pd.astro.it/cinzano/refer/index.htm

14. Cinzano P., 2000a, in Cinzano P., ed., Measuring and Modelling Light Pollution, Mem. Soc. Astron. Ital., 71, 93

15. Lighting and Crime

http://www.asv.org.au/lpoll/lppage.htm

16. International Dark-Sky Association

http://www.darksky.org/resources/quotable.html

ABOUT THE AUTHOR

Dr. Hemant Pathak held positions as Assistant Professor in the department of chemistry, Govt. Indira Gandhi Engineering College, Sagar, MP, India. He had extensive experience in teaching, research and administrative management.

Dr. Pathak received his Ph.D. degree in chemistry from Dr. Hari Singh Gour Central University, Sagar, India and M.Sc. Gold medalist from Jiwaji University, Gwalior. He has published 15 books and more than 50 research papers in reputed International and National journals and received several awards. He is a member of editorial boards and reviewer boards of several international journals and societies. His area of specialization includes Engineering Chemistry and Environmental Pollution management.